CAREERS IN

BIOTECHNOLOGY

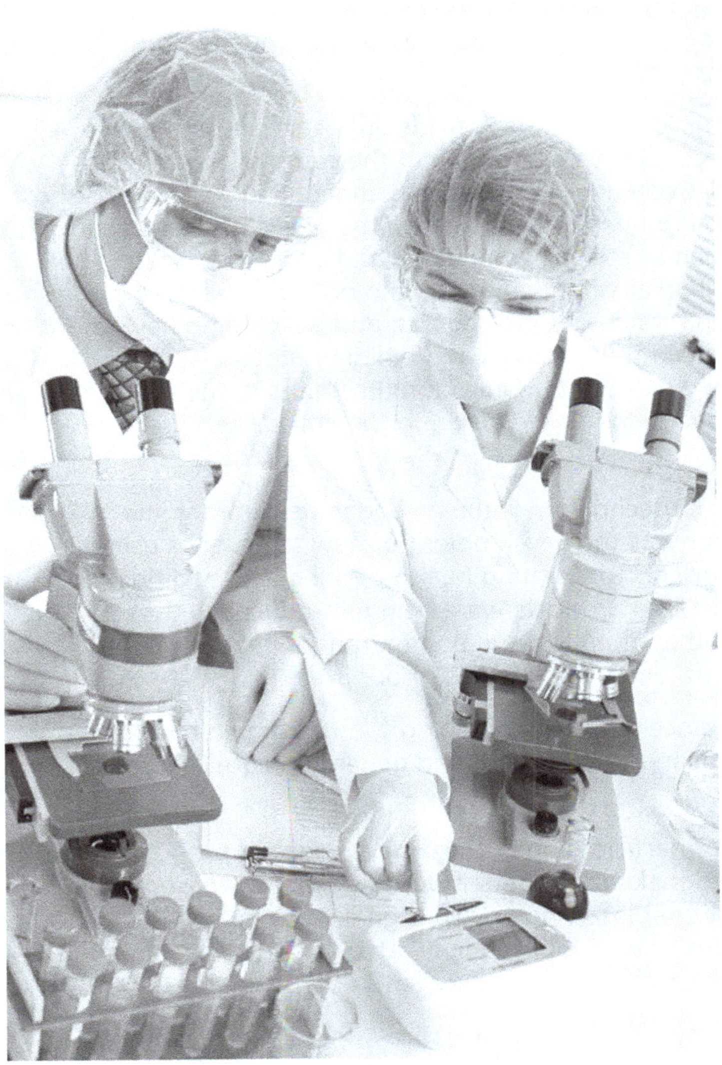

SCIENTISTS ARE PICTURED AS LIVING IN an ivory tower, out of touch with the real world. This might be true of some scientists, but not biotechnologists. These scientists are focused on using biological discoveries in practical ways or, as the United Nations Convention on Biological Diversity noted, biotechnology refers to "any technological application that uses biological systems, living organisms, or derivatives thereof, to make or modify products or processes for specific use."

Biotechnologists work in many fields but especially in agriculture, food science, and medicine. Many of the companies where they work are relatively new and are often engaged in some of the most exciting developments taking place in the world of science and medicine, such as seeking treatments and cures for diseases that attack people, plants, and animals, finding ways to increase the size and nutritional value of food sources, and developing new, more efficient and eco-friendly fuels and building materials. As the Biotechnology Industry Organization (BIO) puts it, the industry's goals are *Healing the World, Feeding the World, and Fueling the World.*

The modern biotechnology industry began to bloom in the 1970s. The key factor was the development of new techniques using DNA to create products for treating health problems. In addition, new biotech-generated products were being introduced with applications that improved the home, transportation, entertainment, and other aspects of our lives.

In the years since the industry took off, it has drawn hundreds of billions of dollars in investment and generated equally large revenues.

The appeal of participating in the development of products that can save lives and make this a better world is certainly reason enough to make a career in biotechnology your goal. Fortunately, this is a growing industry with jobs being created at a fast pace. In addition, earnings in biotechnology are excellent.

There are many different career paths within the biotechnology industry so you have a good chance of finding something that is right for you. Among the different types of occupations are scientists and biotech engineers. These professionals are supported by an army of lab technicians and assistants, sales and marketing people, public relations representatives, and executives who spend a significant part of their time raising funds. A career in biotechnology can also include teaching as an option. Further, there are opportunities for industry advocates. These typically represent a state, regional or even national collection of biotech companies to the public and to the financial community, as well as to elected officials and persons in regulatory agencies, providing them with information about the latest developments that could use financial and legislative support.

A career in biotechnology may have you working in a laboratory or a hospital, an office or classroom. You might also spend time doing field work in an agricultural setting in the US or just about anywhere else in the world.

Most positions in biotechnology require a college education, although laboratory technicians may be trained and certified without having to obtain a college degree. Scientists and engineers as well as teachers typically have graduate degrees, at least a master's but more often a doctorate. Even non-scientists in the industry are typically well educated in the sciences because of their need to communicate with their colleagues and explain the work that is being done to the outside world.

A career in biotechnology demands a good deal of dedication and perseverance. There are more failures than successes coming from the labs. You need to have the kind of confidence that can accept failure and keep going towards your goals. There is the challenge for many in the industry of finding financial support. This requires as much perseverance as the work in the lab.

In the end, the knowledge that you are working towards building a better future for humankind is sufficient motivation for putting your all into your work in biotechnology.

THINGS TO DO NOW

TO BEGIN YOUR CAREER PREPARATION, take as many courses as possible in biology and other sciences, including physics and chemistry, as well as math. These classes will open a window on the work done by scientists that you might be doing yourself someday.

In high school you can also begin to examine different college biotechnology programs. This should help you to decide what area of biotechnology most appeals to you and the type of work you are most attracted to.

You should also participate in extracurricular activities related to biotechnology. These can include school or math science clubs or science fairs. You will benefit from the experience of working with others, something you will be doing throughout your career.

Try to make contact with people in the biotech field. You may do this through volunteering at local companies. Chances are you have a state or regional biotechnical association made up of biotech firms. See if they take on high school interns. If not, see if you can arrange – with the help of one of your science teachers – to have a representative of one of the companies come to speak at your school or just to your class about what's happening in this field. If you can't locate a state or regional association to help you, try BIO, the national organization, at http://bio.org

Another way to get involved is to read the publications in the field, both print and webzines on the Internet. *BIOTech Now* (http://www.biotech-now.org) the online publication of BIO, is a great place to start. BIO also offers a blog on the Internet at biofuelsandclimate.wordpress.com with postings about recent events and developments in the biotech field. Other publications that will get you right into the latest news of the industry include *Bioscience Technology, Biotech International,* and *Biotechnology Progress.* You will find them challenging to read, but they will also give you the best idea of what is important to the people working in the field today.

HISTORY OF THIS CAREER

BIOTECHNOLOGY CAN TRACE ITS ROOTS back to prehistoric times when people discovered that grain, when allowed to get wet and sit for a while, is transformed into a beverage that can be very enjoyable – known today as beer. The primitive people who made this discovery set about trying to understand what had happened so they could make it happen themselves rather than relying on a chance discovery of the next batch. In trying to understand and replicate this organic process, they were the first biotechnologists, doing essentially what their descendants are doing in laboratories around the world today: using biological processes to achieve a specific goal. Along the same lines, the primitive people who discovered that if their grain was mixed with water and left on a hot stone it became something good to eat – what we now know as bread – also set out to understand and reproduce the process so the supply of bread could be controlled.

Civilizations grew up around the utilization of these agriculture-related processes and the need to have an organized growing and harvesting process in order to supply the grains needed. These ancient people who developed agricultural production and tried to maximize the productivity of their lands at the dawn of history were essentially pursuing similar goals as modern biotechnologists.

Through the next several centuries, people around the world continued to observe natural phenomena occurring in the world around them that had nutritional or health benefits and studied these events in order to understand how they worked and if they could be reproduced and controlled, so that the benefits could be drawn upon when needed and not just by accidents of nature.

A great leap forward occurred in the 16th century when the microscope was invented, allowing scientists to grasp what was happening in the natural world at a completely different level. Using this amazing instrument, scientists were able to discover the cellular structure of matter and identify forms of life they previously had no idea existed, such as protozoa and bacteria.

As the microscope became more sophisticated and other technologies developed, further discoveries were made, including the existence of proteins and enzymes, and the structures within cells. By the beginning of the 19th century the term "biology" had come into use, and by the middle of the century scientists like Louis Pasteur were able to understand that yeast, the ingredient key to the creation of bread and beer, was a living organism.

Around the same time, an Austrian monk and scientist named Gregor Mendel made the astonishing discovery that new generations inherited traits from their ancestors through their genes. Once this was understood genetic modifications could be implemented, a major step forward in biotechnology.

Another big step that showed how varied the solutions produced by biotechnology could be came in 1914, when bacteria were used in the UK to treat sewage for the first time. Within five years, in 1919, the word "biotechnology" was used for the first time.

One of the greatest events in biotechnology history came in 1928. The Scottish scientist Alexander Fleming discovered penicillin, the first antibiotic. Over the next few decades, discoveries came fast, including the structure of DNA, the development of cell culturing techniques, the identification of a methodology for gene cloning, and by 1981, the first genetically engineered plant, as well as the first successful cloning of an animal, a mouse.

1982 saw the approval by the Food and Drug Administration of the first biotechnology drug, Humulin, a human insulin drug for the treatment of diabetes produced by genetically engineered bacteria. A year later the first artificial chromosome was synthesized and the year after that the first genetically engineered vaccine was developed.

The late 1980s and early 1990s saw the introduction of several breakthrough biotech products including the first biotech-derived interferon drugs for the treatment of cancer, the first genetically engineered human vaccine, made for the prevention of hepatitis B, a biotech treatment of growth hormone deficiency, and a tomato genetically engineered to resist rotting.

In the decades since, developments in genetic engineering and the understanding cf the human genome have progressed rapidly, with new biotech products emerging frequently, dozens of new companies being launched and the industry's commercial revenues skyrocketing.

WHERE YOU WILL WORK

BIOTECHNOLOGISTS WORK IN LABS, conduct field work in agricultural and industrial settings, work in offices, and also in classrooms. The laboratories differ in the type of research being conducted. There may be a biochemistry orientation or the research may tilt towards biophysics. Labs may be part of private or public companies or they may be located in a college or university setting.

In today's global economy with transnational corporations, a scientist may travel among laboratories located all around the world, all belonging to the same company or to a consortium of companies and universities all working together on a single project, but each focusing on a distinct aspect.

Field work may also be a global affair, with scientists and technicians traveling the world to see, for example, how experimental crops are doing on test farms on nearly every continent and below the surface of lakes, rivers, and oceans.

When not in the lab or the field, biotechnologists may be in their offices writing reports on their work and preparing grant applications. They may follow up by traveling to conferences, where they meet with other biotechnologists to provide each other with updates on the work in which they have been engaged. Travel may also be a follow-up to the grant application process, for meetings with potential funders.

If you choose to participate in biotechnology as a teacher you will spend much of your time in a classroom. You may also be involved in research and split your time between classroom and lab.

THE WORK YOU WILL DO

BIOTECHNOLOGY SCIENTISTS AND TECHNICIANS typically specialize in one area – biofuels, agriculture, or one of the many areas within healthcare. Within that specialty, their focus becomes even more narrow, with many scientists dedicated to the study of a single type of natural function or organism with the aim of first understanding all they can about it and then how it can be used or altered so that it contributes to improving crop yields, providing cleaner fuel, treating a disease, etc.

As a scientist or technician, your work will be done either in the lab or in the field, which can mean an agricultural setting, a healthcare facility, or a factory, depending on the area of focus. The actual work will vary, but in the broadest terms you will be using biological materials, breaking them down and rebuilding them chemically, combining them with other materials, or trying to isolate a specific component within a biological substance. You may be seeking a higher yield crop, a cure for cancer, or a fuel that doesn't contribute to global warming. The work may be tedious as well as exciting, requiring what may seem like the endless repetition of tests and measurements, and ongoing recording of data from your experiments.

The equipment you will use can include microscopes, slides, Petri dishes, grafting tools, and measuring devices and, of course, computers. The great progress biotechnology has made in recent years is in part the result of advances in computer technology, which allows scientists and technicians to model different solutions before they actually do a test. This saves the time and expense of running tests that have no chance of success. The use of computers in biotechnology has been given its own identity and specialty called bioinformatics.

Biotechnologists tend to work in either biochemistry or biophysics. Biochemists study the chemical composition of living things. Biophysicists study the role of physics in the processes of living organisms.

Some of the subject areas that function within biotechnology are microbiology, physiology, botany, zoology, ecology, and agricultural and food science. As a scientist and technician your work may take you out of the lab to field sites where instead of operating a microscope, you may be handling animals, crops, and engines. You may also spend time in healthcare settings working directly with human subjects. Your work may take you under water, as science looks at the makeup of sea creatures and plants for new elements with the potential to help life on land.

Part of your time as a scientist will be spent writing or contributing to grant proposals to obtain funding. Colleges and universities, private foundations, and federal government agencies, such as the National Institutes of Health and the National Science Foundation, contribute to the support of scientists whose research proposals are determined to be financially feasible and to have the potential to advance new ideas.

While biotech scientists may conduct their analysis of the research results in their offices, in the lab and in the field they work with technicians. Technicians in biotech labs and biotech field operations are responsible for the setup, operation and maintenance of instruments and other equipment, and they typically have responsibility for monitoring and recording data as research projects progress. Technicians have, in recent years, become increasingly responsible for developing lab procedures and protocols for the specific projects they are working on.

Biotechnology research usually requires elaborate safety precautions. There are always concerns that living organisms like microbes and viruses must be contained and not be allowed to spread to the outside world. In addition, there are internal safety precautions to protect the scientists and technicians working with these materials. Many jobs in the biotechnology industry are specifically about monitoring and enforcing safety standards in labs, test fields, factories, and transportation. These jobs tend to be within government agencies like the Food and Drug Administration and the Environmental Protection Agency. There are also private companies that a research firm might employ to ensure it is in compliance with the regulations within which it operates. The people who monitor and enforce regulations are, like the people whose work they are observing, scientists, technicians, and managers.

Related to the work of safety inspectors is that of environmental scientists and technicians who have responsibility for waste management operations, inventory control, and the management of hazardous materials.

Biotech research operations often have several scientists and technicians participating in a team effort. Heading research operations are managers who are often scientists themselves but can also be engineers or administrators who don't necessarily engage directly in the research but whose responsibility is to see that projects stay on schedule and on budget. These managers are knowledgeable about the science involved in the research and will probably have participated in the design of the project.

In addition, a manager would work with the lead scientists on funding requests that preceded the project's launch. On the other side of the project's life, the managers will again work with the scientists on preparing the presentation of the results.

A manager may be responsible for several projects at the same time including those being conducted within the company and by outside contract organizations. Besides time and budget concerns, managers also bear the ultimate responsibility for safety and for regulatory compliance.

Managers may also have staffing responsibilities, including hiring, firing and evaluations. In addition to project responsibilities, managers may train new staff members in the standard operating procedures and methodologies followed by the company, and compliance of the work site with industry regulations.

Managers spend a considerable amount of their time in communications mode. In addition to the team or teams they are supervising, they also have to report to senior management and to regulatory authorities. They are also in frequent contact with their colleagues in marketing and production as well as with outside contractors and suppliers.

Some managers are also engineers, which means they may be involved in production when a research project results in the successful development of a new product or process. This could include the design of production equipment, the implementation of a new process, or the creation of new packaging or delivery systems.

The academic world is an important part of biotechnology. Universities serve as research partners with individual companies or as components within a consortium of private companies, government agencies, and other universities. While some teachers in biotechnology are solely teachers, most split their time between the classroom and the lab.

Besides scientists, technicians, managers, executives, and academics, the biotech industry workforce includes a number of other key positions. Among the most necessary careers are those that involve patents, because this is an industry that is focused on bringing new products to market. Patent attorneys and patent administrators are important contributors to the process, preparing, filing, and prosecuting patent applications. It is important that these professionals have an understanding of the biotechnology sciences so that their documents are accurate in describing a new product or process. Patent administrators are responsible for all aspects of filing and prosecution of patent applications with the US patent office and with foreign patent offices.

Another important position in this broad industry is biostatistician, a position that requires math skills, advanced computer skills, and an understanding of the sciences involved. Biostatisticians collaborate with research scientists and technicians to design studies, providing direction on survey structure, sample size, data collection and related topics.

Technical writers are also important contributors. They are responsible for preparing reports that simplify the science involved in biotech projects for non-technical readers who may include funding sources, government agencies, end users, the press and even the public. Technical writers might find themselves working closely with communications, public relations and media staff as well as with the scientists whose work they are documenting.

BIOTECHNOLOGISTS TALK ABOUT THEIR CAREERS

I Am Currently President & CEO of BayBio, Northern California's Life Science Industry Organization

"BayBio provides a variety of educational and networking programs to life science professionals, advocates in Washington DC, the state capital Sacramento, and locally, for a positive business climate to promote innovation, support science education, and help bio-tech entrepreneurs and startup companies. In my role, I am a spokesperson for the industry with the media, legislators, regulatory authorities and company executives.

I have always been fascinated by the ability to understand and harness biology to impact disease. As a management consultant working across a wide variety of industries, I have found that biotech offers a unique opportunity to work with very smart, innovative, entrepreneurial people, satisfy my intellectual curiosity and have significant impact on patients.

To succeed in biotechnology you should have a passion for science, whether you are in a scientific or business role. You should also have an entrepreneurial spirit, meaning you are willing to take risks, think creatively and explore new ideas. You will also need a real desire to make a difference in people's lives.

The greatest rewards of your work come from knowing that you are helping people. Working with very bright, creative, dedicated people is a great plus. Staying current on the newest technologies and pushing the boundaries of business and science are very rewarding.

One hard part of the job is gaining access to capital to fund innovation. The reason funding can be hard to find is the inherent risk and unpredictability of biology. Science is usually well ahead of politics, and it is a big challenge to educate legislators and regulators as science advances rapidly.

My advice is to take as many science courses as possible. There are many interesting career options available, many outside the lab, but having a foundation in science is important for them all – even for business positions. Ask your teachers to sign up for National Lab Network at www.bio-community.org. At this website you can request mentors, company tours, internships and career talks."

I Am the Environmental Monitor for a Pharmaceutical Company

"I make sure that we are in compliance with all the rules and regulations set down by the Food and Drug Administration. We get inspected from time to time and if we are not crossing all the t's and dotting all the i's we can be shut down and/or fined. Lost time, lost credibility and lost revenues and funding are all looming if we slip up.

Environmental monitoring requires reviewing data generated by equipment in our labs and manufacturing facilities, as well as by human observation. I also conduct trend analysis to make sure we are heading in the right direction in terms of our standards, or at least to make sure we aren't getting off track.

If anything does go wrong, I'm the one who leads the investigation. I'm a little bit like Sherlock Holmes, going around with a magnifying glass looking for clues. If I uncover a problem, I write up the recommendations to avoid the same thing ever happening again.

Since we are always changing our methodologies, I work with the production teams and the lab directors to make sure these changes don't take us out of the regulatory parameters. My input can help determine the scope of the changes so that we don't have to go back after the fact to make corrections. What we want is to come up with standard operation procedures that are clear and effective.

I prepare a quarterly report that includes the relevant statistics and a summary noting any difficulties we encountered and how we dealt with them. I also do an annual report that looks back and discusses any changes we are planning for the following year. I contribute to the budget planning, too.

Anyone who wants to do a job like this needs to be very detail-oriented and extremely persistent. It also requires some courage as you have to tell people things they don't want to hear, like, we have to redo a whole project or invest a substantial amount of money to correct a mistake. Of course, you must love biology, but you should take all types of science and computer classes as a way to prepare for working in biotechnology as a quality assurance manager or in any other capacity. Try to get intern work if you can. In the end, this is a very rewarding field – everything we do is aimed at helping make life better for everyone."

My Current Position is President and CEO of BioOhio, the Nonprofit Bioscience Organization for the State of Ohio

"Over my career, I have been head of worldwide R&D (research and development) for a multi-national corporation. I have started five companies (four in biotech) in both Europe and the US, and I have traveled the globe pursuing my research and the advancement of the biosciences.

I became involved in biotechnology through an interest in medicine that focused on microbiology. Once I discovered how interesting microbes were, I was hooked. In college and then graduate school my interests expanded to cancer research.

I would say the greatest skills and qualities a person needs to succeed in biotech (in addition to a good education) is an unending curiosity about how the natural world works, and the drive to never be satisfied with the status quo.

The greatest rewards from my work are a sense of satisfaction that I am advancing science and civilization in some small way, and that the products and services I have helped launch make a difference in people's lives.

Beyond the rigorous education, the greatest challenges are coping with failure and persevering (there is a 90 percent failure rate with any kind of research), and learning when to work with the team and when it is better to strike out on your own to get something done your own way. I have found that both independence and teamwork are necessary elements of success – the key is knowing when to use them.

Most states have active education programs in biotechnology at many levels – interested students should seek them out either through their schools, through an organization like ours, or nationally through BIO (the biotechnology industry organization). I cannot speak for all of the states but here in Ohio we have programs like Project Lead the Way and the Third Frontier Internship program sponsored by the state."

I Am a Senior Account Manager for a Global Company Making Oncology Drugs

"I have an assigned set of customers that includes hospitals, group purchasing organizations, and other healthcare operations, and I work with them to define their therapeutic objectives. We want to have a long-term relationship with our customers so it is important for me to be able to understand exactly what their needs are in terms of their patient base, their treatment plans and so on. It also means that I need to report back to our scientists and my managers about what the customers say they want from us. In a way, I feel like I work for the customers as much as for my company.

There was so much to learn when I first started. Besides knowing about the products and our company and our customers, I had to know who our key competitors were and all about their products so I could explain the differences. I also had to learn a load of technical terms about the treatments, the conditions, insurance coverage, government regulations, and much more. Even though I had a basic education in biology, this was pretty overwhelming at first. It was essential to master it all; as I need to provide the best advice that I can to our customers and be able to make clear exactly what I mean.

I'm always impressed about how smart all the people I work with are and how much they know, not only about our products but also the business side of things. If you are looking at a career in biotechnology you had better be ready to really apply yourself and know that you will always have to keep learning new things as new drugs are developed. This is true even if you are just going to be working in sales.

You should also enjoy traveling. Obviously I have to travel constantly because I'm almost always on the road visiting my customers. I see our scientists traveling, too. Sometimes they come with me to call on some of our larger customers, but they are always going off to meetings and conferences. I personally like to travel but if you don't this could be a real problem."

I Am the President of the Washington Biotechnology and Biomedical Association (WBBA)

"Our mission is to help grow life sciences in Washington state. After finishing my master's degrees in audiology and getting an MBA, I was trying to get into the hearing aid industry, but there were no jobs. I applied for a job as a pharmaceutical rep for ER Squibb & Sons and was hired. I did well, moved up and was recruited by Centocor in 1989. I have been in the field ever since.

The rewards of being in this industry are great. I have worked for and started companies that have brought treatments to patients that have saved countless lives. I have had patients and their family members thank me for saving their life. It's very, very rewarding.

A good science background, especially in biology, and a good math background, are essential to success in biotechnology, no matter what kind of work you'll be doing. If you are interested, reach out to local companies and research institutions. Many of them have internships or events to help increase awareness. Also, reach out to local and state trade associations."

I Am the President and CEO of the Colorado Bioscience Association.

"I manage a staff of four and provide education, marketing and advocacy to over 350 pharmaceutical, biotechnology, medtech (medical technology), and agbio (agricultural biology) companies in Colorado.

Our team works together to create valuable services to these companies. I also do fundraising and recruit members since it is what pays for our operations. In my job I spend most of my time advocating for bioscience companies at the state legislature and consult with our Congressional delegation.

My greatest reward is working with exciting companies every day that are developing drugs and devices that help to improve people's lives. It takes on average 10 to 15 years and $1 billion to develop a new drug, so there are many financing challenges to help small emerging companies succeed. If they don't, then many amazing ideas may never come to fruition. We could be losing potential cures to diseases or new therapies.

A young person interested in biotechnology should have a strong science and math background, personal drive, curiosity, a good work ethic, and the ability to work with others. I would strongly suggest that when possible students should take advantage of volunteering at a company or a hospital/healthcare facility to get some experience dealing with people and to see the hundreds of different jobs that they can pursue in biotechnology – whether in research or as a healthcare worker.

One important thing to remember is that you can do IT (information technology), accounting, business management and other jobs in the bioscience industry since they are all vital to the industry. Not everyone has to have a PhD or an MD."

I Work for a Company Developing Alternative Fuels From Biomass Materials

"A major part of my job is keeping track of all the developments taking place in the field – including at other private companies and also at government and university labs. We're not trying to steal anyone's ideas, just making sure we aren't reinventing the wheel. There actually is a lot of sharing that takes place in the field of biofuels, at least once you have your patent applications in place.

Because I have degrees in both business and chemistry I find that my responsibilities tend to include developing our business plans, determining costs for equipment and materials, and staffing needs. I am often involved with other companies working out plans for joint ventures to help reduce the costs we all face.

Working in biofuels or any biotechnology field requires a willingness to be open to new ideas. No one has all the answers and we are all learning from each other all the time. This is true on the business end as well as for the biologists and chemists involved in the field.

You also need to be patient. More often than not the research you are working on is not going to be successful. In addition, funding for new projects is not always easy to come by, and it takes intense and constant work to make it happen. You need to keep a positive attitude.

We are all chasing a big prize, trying to come up with a breakthrough product that will be a game changer for the fuel industry. Making that breakthrough is the big reward, but I feel like having the opportunity to even try is also rewarding. The hard part is trying to keep that in mind when things are not going well."

PERSONAL QUALIFICATIONS

BIOTECHNOLOGY IS A HUGE AND well-established industry yet it still retains a pioneering feeling thanks to all the startups that continue to appear. To fit in, you should be passionate about science, about making discoveries and breaking new ground, and about creating a new and better world. This feeling is important whether you are in a scientific or business role. You must be willing to think creatively and take chances.

You must be grounded in a strong science and math background. This is important for both scientists and non-scientific industry personnel. The reverse is also true: if you are a scientist in the biotechnology industry you should have a good business sense. You may be called upon to participate in business decisions, trying to determine the value of new products, cost of production, best sources for materials, or other factors that will be significant to business decisions that must be made. You will need to be familiar with regulatory issues. and marketing and management techniques.

Being a self-starter is important, and is a necessary ingredient in this entrepreneurial environment. Nearly as important is the ability to work well with others, as you most certainly will whether in the lab, the office, the field, or at conferences and in business meetings. It is also of great value to be able to communicate clearly and concisely, both speaking and writing.

For the adventure of field work, you will need to be in good health and physical shape, especially if your field work takes you to challenging environments around the world.

Another quality you will need in any type of career in this field is patience. Projects may take years to reach fruition and finding financial support can seem like an endless chore. Success is very often tied to endurance.

ATTRACTIVE FEATURES

PROFESSIONALS IN THIS CAREER LOVE the personal satisfaction that comes from working on products and services that help improve people's lives or, as one person said, "A sense of satisfaction that I am advancing science and civilization in some small way." When a product or service – no matter how small the difference it makes – gets to the marketplace and becomes a success, the satisfaction level goes through the roof.

This field offers the opportunity to work with people as dedicated as yourself, and who are also in love with the opportunity to be creative as well as being of service. These include your immediate colleagues as well as the colleagues you get to share with at conferences and meetings.

The opportunity to travel is another attractive feature for many in the biotechnology industry. Projects often feature multinational participation, so that scientists and businesspeople often have to visit cooperating sites around the world. In addition, there are conferences and meetings that require travel. For those who desire excitement as well as the challenges of science, the chance to work in an exotic setting is very appealing.

One other attractive feature of a career in biotechnology, especially for scientists and technicians, is that it allows contact with the most advanced *development.*

UNATTRACTIVE FEATURES

NUMBER ONE REQUIREMENT OF THIS career is patience. That may be the number two and number three requirement, as well.

Endless patience must be paired with a thick skin in order to endure the failure and rejection that are common to the biotechnology experience. As one participant noted, research is 90 percent failure. You have to be quite the optimist with a great degree of perseverance, in order to keep going, picking up the pieces and starting over when your efforts come up short.

If you are inclined to work alone, or at least with the minimum of interference from others, then a career in biotechnology may not be right for you. Teamwork is the norm, where many breakthroughs have been the result of several contributors working together.

In a time of economic downtown, the struggle to raise funding has become a major aggravation. The frustration that comes when a request is rejected is akin to that which takes place in the lab when an experiment does not get the hoped-for results. Not only is there the problem of having to keep looking, but also the awareness that potential cures to diseases or new therapies may not be found as a result of the lack of funding.

EDUCATION AND TRAINING

THE EDUCATION YOU WILL NEED WILL depend on the type of position and the particular segment of biotechnology you work in.

With a bachelor's degree you can begin working in the biotech industry at an entry level position as a research assistant. Earning a master's degree, requiring an additional two years of study, will create new opportunities with more responsibilities as well as higher earnings. Scientists will likely need to spend an additional two to five years pursuing a doctoral degree in order to be truly competitive in the job market.

Courses for students aiming to be scientists should include, in addition to the core chemistry and biology, mathematics, physics, engineering, and computer science. Those interested in environmental science should take courses in environmental studies and become knowledgeable about applicable legislation and regulations. Also of value are business courses and English courses that can help prepare you for writing grant proposals and working with management to determine product strategies.

Selecting a school that is in line with your goals is extremely important. Most colleges and universities offer a variety of specialized programs. However, a few schools have begun to offer programs that use the umbrella term biotechnology because they take an interdisciplinary approach. For example, Colorado State University offers a Biotechnology Interdisciplinary Studies Program (BISP) in which students take core courses in biochemistry, microbiology, bioprocess engineering, and biotechnology, and electives tailored to the individual student's interests such as Principles of Animal Breeding, Cell Biology, Plant Physiology, and Food Microbiology.

Another school that serves as a model for where biotechnology education is heading is Washington State University. WSU offers a bachelor's degree in biotechnology and also a highly interdisciplinary doctoral program in biotechnology. The school is highly regarded for its internationally recognized Center for Integrated Biotechnology. The Center facilitates and expands the opportunities for students studying biotechnology with training and research programs. The Center is home to several core laboratories, including genomics, proteomics (the study of proteins in living organisms), molecular biology, bioinformatics, plant transformation, and laser microdissection.

WSU also offers the NIH (National Institutes of Health) Protein Biotechnology Training Program which features interdisciplinary training intended to enhance the experience of graduate students. Students in the program are described as trainees and are recruited from related departments such as chemistry, bioengineering, molecular biosciences, pharmacology/toxicology, and veterinary microbiology and pathology, and are supported with grants from the National Institute of General Medical Sciences, part of the

National Institutes of Health.

Technicians will almost certainly need at least an associate degree. However, there are non-degree certificate programs in applied science or science-related technology that may be acceptable to employers. As with a bachelor's degree, the associate program courses covered in a two-year program include classes in biology, life sciences, chemistry, physics, mathematics and computer science. Those who are inclined toward working in the biofuels segment might also take courses in environmental studies.

Some community colleges that offer programs in biotechnology may also have internship programs that place students with local biotech companies while they pursue their classes. The same may be the case at technical institutes which are usually more focused on training and less on providing a general liberal arts education.

The importance of including computer courses continues to grow. Computers are used in all aspects of biotech work, from running lab equipment to projecting research results, or simulating biological processes.

Your interest in biotechnology does not mean you have to become a scientist. This still-emerging industry requires the key participation of entrepreneurs and business leaders who have pursued an MBA (Master of Business Administration degree) instead of a PhD in biology. However, while not required, a second degree in science or at least a minor in one of the sciences will be valuable for non-scientific personnel, adding to the credibility of executives, sales personnel, media relations representatives, and other non-science positions. There are many other occupations within biotechnology such as quality control, quality assurance, information technology, human resources, facilities and infrastructure maintenance and manufacturing where a science background will be helpful.

Duel master's degree programs in business and biotechnology are beginning to appear in the US. The University of Pennsylvania, for example, offers what it describes as "one of the deepest immersion experiences in the fields of biotechnology and business" through its dual-degree program. The program prepares individuals to meet the challenges and opportunities in the biotech industry through "in-depth training in recombinant DNA technology, biopharmaceutical/medical devices, or bioinformatics with a grounding in management."

The MBA/Biotechnology Program offers specialized courses in genomics, drug discovery and development, pharmaceutical manufacturing, biomolecular engineering, and biotechnology entrepreneurship. In addition, the program includes a biotechnology seminar where students "can meet experts from venture capital finance, biotech startups, and pharmaceutical houses." Penn notes that its location in Philadelphia gives students easy access to one of the top biotechnology hubs in the US "where over 70 biotechnology startups exist within a 50-mile radius of the campus."

EARNINGS

SALARIES FOR SCIENTISTS WORKING for biotechnology companies fall within a wide range depending on the age, size, and success of the company where they work. The same is true for technicians as well as the front office staff who market, sell, raise funds, and deal with government agencies and competitors in the field. Experience and knowledge also factor into how a salary is determined.

Salaries for scientists fall in a range at the low end between $40,000 and $50,000 per year, up to between $130,000 and $150,000. In general, the salary of an experienced biologist or biochemist will be in the $80,000 to $100,000 range.

Pharmaceutical companies and private research firms offer the highest salaries. They can also offer substantial bonuses and, startups often offer stock options with the potential of very high profit. Government agencies can offer salaries that are competitive up to a point, with the added benefit of job security. Nonprofit organizations and academia will in general offer scientists and technicians the lowest salaries.

Median annual wages for business managers in the biotech industry are upwards of $100,000 per year. The pharmaceutical industry offers the highest salaries, with private research firms not far behind. Median salaries for management level personnel in government agencies that conduct biotech research or interact with the industry are around $80,000. As with the scientists, managers in private industry have benefits such as stock options and bonuses, while government positions offer more security.

Biotechnology technicians' earnings are in a range of $35,000 to $55,000 depending on their particular specialty and the type of organization where they are employed. Technicians in private industry may receive bonuses and stock options in addition to their salaries. Those who work in government agencies have more job security although their salaries may average lower than those in private industry.

OUTLOOK

THE BIOTECHNOLOGY INDUSTRY HAS shown constant growth over the past decade, with employment just about double where it was at the start of the period. The consensus is that this growth will continue, although more slowly given the pace of the economic recovery taking place in the US and other parts of the world which has seen a reluctance to invest the venture capital that is essential to development.

In addition, government support for biotechnology research is likely to slow as politicians at the federal and state level look to cut spending in their efforts to balance their budgets. Further, a high percentage of biotechnologists work directly for the government in the Departments of Agriculture, Energy and Interior. Many also work at the National Institutes of Health. If there are cutbacks in government spending they may include reducing agency staffs.

Why the optimism about job growth, then? For one thing, the pharmaceutical industry, a major employer of biotechnologists, is not hurting when it comes to profits, and reinvesting in research is a core component of the industry, promising good opportunities for future biotechnologists.

The majority of biotechnologists work in the private sector, and while the number of new startups may have slowed, it has not disappeared. As the economic growth continues and expands, biotechnology will be among the industries likely to lead in the recovery. There is an overall recognition that biotechnology is a 21st century industry, and the private and public sectors will continue to cooperate in the effort to see that the industry in the US thrives.

In addition to the jobs within the industry, in government agencies and in academia, there may be considerable growth in other industries. Biotechnology professionals will be employed by many companies which want to stay aware of what the developments in biotech might mean for them in terms of new building materials, new energy sources, healthcare breakthroughs, and other advances. The interest may be for investment opportunities, as well.

Another source of opportunities for new graduates will be the retirement of many older industry participants, especially in government positions. In general, new jobs will be created at a pace that, even without the retirements, would be considered faster than most professional careers. The competition for new openings will be tough, and those with master's and doctoral degrees will have the best chances of finding employment in the lab and in the front office.

GETTING STARTED

WHILE STILL IN COLLEGE IT IS POSSIBLE TO BEGIN YOUR career in biotechnology by finding part-time or internship opportunities with companies participating in the industry. Your own professors are the best place to begin searching for these types of positions as many biotechnologists in academia have connections to the industry. In addition, many professors have budgets that allow them to hire assistants to help with their research.

You can also contact your state or regional biotech organization for information on internship opportunities. These professional organizations may also offer forums and blogs where students and professionals in the field exchange information and discuss ideas and current developments. There are blogs that offer interesting insights into the current state of the industry such as:

Bioinformatics
http://barf.jcowboy.org

BioInfo
http://bioinfblog.blogspot.com

An organization expressly for students that is worth investigating is Student Pugwash USA (www.spusa.org). SPUSA aims to promote social responsibility in science and technology. Members are encouraged "to make social responsibility a guiding focus of their academic and professional endeavors."

It's also important to keep up on industry developments through publications, both hard copy and online. Monitoring the general business press is important, too. These sources can help you to understand the directions being taken in biotechnology and in the context of the larger economy. Keeping abreast of the global political scene is also very important as biotechnology developments are often subject to intense political pressures.

Go to a conference, if possible. Interesting and valuable conferences and meetings on all aspects of biotechnology are held around the world with many of the events in the US. Most take place in Washington, DC, probably for the proximity to government funding sources. Other significant meetings are held in Boston, Ft. Lauderdale, San Diego, San Francisco, and in Puerto Rico.

A schedule with links to upcoming events can be found at www.biotechnologymeetings.com

All of the contacts you'll be making, as well as the reading you'll be doing, and, of course, your classroom experience, will help you to determine where, within the wide world of biotechnology, you want to devote yourself, especially if you are going on to work as a scientist. Technicians and business executives can have similar choices, as to whether, for example, they want to focus on biotechnology efforts that address environmental issues, or ones that look to health, agricultural, or biofuels issues. The more outreach you can achieve during this period, the better informed you will be when the time comes to make this decision.

ORGANIZATIONS

In addition to the national and international organizations listed below, many states have biotech organizations, for example the Illinois Biotechnology Industry Organization at www.ibio.org and the North Carolina Biotechnology Center www.ncbiotech.org

■ **American Association for the Advancement of Science www.aaas.org**

■ **American Academy of Health Physics http://hps1.org/aahp**

■ **American Association of Physicists in Medicine www.aapm.org**

■ **American College of Medical Physics www.acmp.org**

■ **BioIndustry Association www.bioindustry.org**

■ **BIOTECanada www.biotech.ca/en/default.aspx**

■ **Biotechnology Industry Organization www.bio.org**

■ **Biotechnology Information Institute www.bioinfo.com**

■ **Biotechnology Institute www.biotechinstitute.org**

■ **Center for Advanced Biotechnology and Medicine www2.cabm.rutgers.edu**

■ **Center for Biotechnology**
www.biotech.sunysb.edu/index.html

■ **Council for Biotechnology Information**
www.whybiotech.com

■ **Food Biotechnology Communications Network**
www.foodbiotech.org

■ **National Center for Biotechnology Information**
www.ncbi.nlm.nih.gov

■ **National Agricultural Library**
www.nal.usda.gov

■ **Parenteral Drug Association**
www.pda.org

■ **Society of Bioprocessing Professionals**
http://sbpeducation.com

PERIODICALS AND WEBSITES

■ **American Biotechnology Laboratory Magazine**
www.americanbiotechnologylaboratory.com

■ **American Laboratory**
www.americanlaboratory.com

■ **BioInfo**
http://bioinfblog.blogspot.com

■ **Bioinformatics**
http://barf.jcowboy.org

■ **BioOptics World**
www.optoiq.com/index/biophotonics.html

■ **Biophotonics**
www.photonics.com

■ **BioProcess International Magazine**
www.bioprocessintl.com

■ **Biotechniques**
www.biotechniques.com

■ **Lab Animal Magazine**
www.labanimal.com

■ **Life Science Leader**
www.lifescienceleader.com

■ **Nature Biotechnology**
http://www.nature.com/nbt/index.html

■ **Nature Methods Journal**
www.nature.com/nmeth

■ **Photonics Spectra**
www.photonics.com

■ **Signals Magazine**
www.signalsmag.com

■ **Your World: Biotechnology & You**
www.biotechinstitute.org/your-world-magazine